Primary Ecology Series

Squirmy Wormy
Composters

Bobbie Kalman & Janine Schaub

 Crabtree Publishing Company

The Primary Ecology Series

To teachers Jackie Stafford & Cindee Karnick-Davison

Writing team
Bobbie Kalman
Janine Schaub

Editor-in-chief
Bobbie Kalman

Editor
Shelagh Wallace

Design and computer layout
Antoinette "Cookie" DeBiasi

Cover mechanicals
Diane Coderre

Type output
Lincoln Graphics

Photo processing
Ray J. Kunnapuu/Heritage Images

Color separations
ISCOA

Printer
Lake Book Manufacturing

"The pixie that put the worms to sleep"
written by Chris Taylor

Artwork and cover design:
Antoinette "Cookie" DeBiasi

Photographs:
Bobbie Kalman and Antoinette DeBiasi:
pages 5-17, 23-29
Albert Eggen: page 4
Janine Schaub: pages 20, 31
Crabtree Publishing Company made every effort to secure model releases.

Special thanks to: Jackie Stafford, Cindee Karnick-Davison, Dean Steele, the students of Parliament Oak Elementary School, the students of Elmlea Junior School, Albert Eggen, Melissa and Katie Drohan, Victorian Eady, Justin Pepe, Mark Jones, Justin Hope, Richard Wilkinson, Rebecca Prewitt, Ashley Caldwell, Ian Hawksbee, Callen and Eloise Adams, and Jamie Kennedy (boy on cover)

Published by
Crabtree Publishing Company

350 Fifth Ave.	360 York Road, R.R.4	73 Lime Walk
Suite 3308	Niagara-on-the-Lake	Headington
New York	Ontario, Canada	Oxford OX3 7AD
N.Y. 10118	L0S 1J0	United Kingdom

Cataloguing in Publication Data
Kalman, Bobbie, 1947-
    Squirmy wormy composters

(The Primary ecology series)
Includes index.
ISBN 0-86505-555-6 (library bound) ISBN 0-86505-581-5 (pbk.)

1. Worms - Juvenile literature.
2. Compost - Juvenile literature.
3. Soil ecology - Juvenile literature.
I. Schaub, Janine. II. Title. III. Series

QL386.K35 1992   j595.1

Contents

Perfect pets

Have you ever asked your parents for a pet? What would they say if you asked them for 1,500 pets that never had to be taken for walks?

Worms are perfect pets. They are gentle, quiet, and they never need to visit the veterinarian. They have no smell, and they do not need to be cleaned. Best of all, worms are good for the environment because they eat garbage. Worms can help you reduce your household or school garbage by as much as a quarter, and the garbage you feed your worms will not end up at a landfill site. Many landfill sites are now full or close to being full!

You can have thousands of worms like these!

Worms can help you cut down on your home or school garbage. They will teach you a lot about nature and the environment. Having worms in your home or classroom is fun, too!

Recycling wonders

Worms are great recyclers! They eat plant and animal leftovers such as apple cores and eggshells. Not only do worms eat garbage, but the waste material they leave behind can be used on your house or garden plants. It is full of the nutrition that plants need for growing and staying healthy.

What is vermi-composting?

When you feed leftover food to worms, you are **vermi-composting**. Whether you do it at home or at school, vermi-composting is one of the most important ways you can help the environment. Watching worms in action will also teach you a great deal about how nature works.

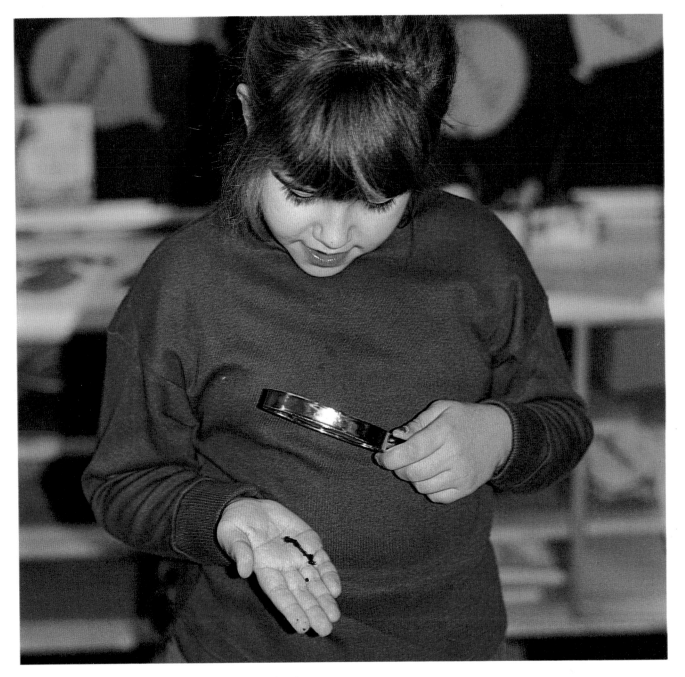

Holding a worm

The first time you hold a worm, it may tickle your hand as it squiggles over your fingers. This feeling might surprise you, so be careful not to drop or injure your squirmy friend. Make your hand into a little cup so that you give the worm space to move around but not enough room to escape. If you are outdoors, move to a shady spot.

Worms do not like being in direct sunlight. Put the worm down on a hard surface and watch what it does. Which end is its head? How does it move? What does it want to do? If you have a magnifying glass, look at the worm through the lens. What does this close-up view show you that you could not see before?

The children on this page have decided to get to know their worms in their own way. They placed them on their arms, noses, cheeks, and ears. One boy even put a worm on his tongue! (See page 29.) We do not recommend that you play with worms because worms don't like being handled, but these pictures demonstrate that you need not feel squeamish about touching them.

The girls (below right) have adult and baby worms, cocoons, and castings on their plate.

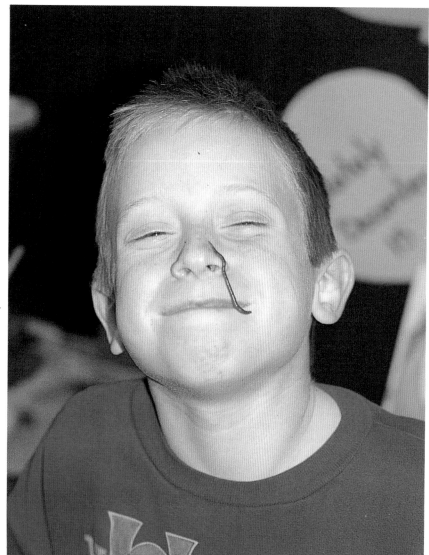

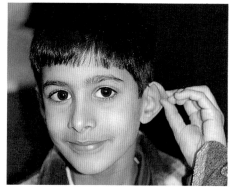

Welcoming worms

The worms we talk about in this book are called Red Wigglers. There are different kinds of worms that can be used for vermi-composting, but Red Wigglers are one of the best types. These lively red worms are good vermi-composters for the following four reasons:

1. Red Wigglers do not mind living in crowded bins.
2. Red Wigglers have lots of babies.
3. Red Wigglers eat heaps of kitchen scraps.
4. Red Wigglers are easy to buy from garden centers, mail-order worm farms, or fish-bait companies.

A home for your worms

A cool, damp, and dark tunnel in the ground is a worm's **habitat**. A habitat is a place where plants or animals naturally live. A worm digs its way through the soil, looking for food. Animals and plants that die and decay provide the worm with the food that it needs to survive.

When you welcome worms into your home or classroom, you must make sure that you provide them with a habitat that is similar to their natural home. The worm's new habitat will be a bin half-filled with a mixture of earth and shredded paper, called **bedding**.

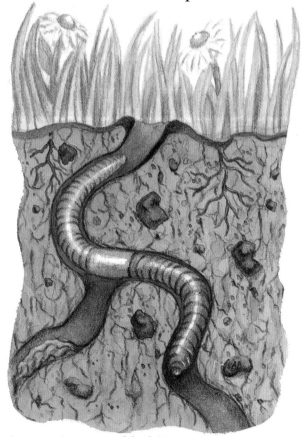

A worm's natural habitat

The worm bin—the worm's new habitat

Not too hot, not too cold!

The perfect temperature for your squirmy pets is between 13 and 25°C (55-77°F). At this temperature, worms eat well and produce lots of babies. If worms are too hot or too cold, they may not eat all the scraps that your family feeds them. They may also find it difficult to **reproduce**, or have baby worms. If the temperature in the worm bin rises to more than 29°C (84°F), the worms can die. Worms will also die if they are allowed to freeze.

Worms need air to breathe, just as people do. A worm bin must be placed in a spot with plenty of air moving around it. The bin must have air holes in its lid. Your worm-bin cover should also keep out light. Worms are sensitive to light and will move away from it. More than a few minutes of sunlight can be harmful to worms.

Starting a vermi-composter

Anyone can start a worm bin in an apartment, house, or school. By following the steps on these two pages, you can easily set up a vermi-composter!

It does not matter if you buy or make a worm bin. Worms are happy living in any kind of bin as long as it is big enough to hold all the worms you need. The bin should have air holes in the lid and holes in the bottom through which water can drain out. It should sit on a tray to catch the water that drains out of the bedding. The bottom of the bin should be covered with fine nylon mesh to stop the worms from escaping through the holes.

Place a fine nylon mesh on the bottom of the bin to prevent the worms from crawling out.

A family of four to six members needs a large worm bin. Your family's bin should measure approximately 30 centimeters high, 60 centimeters wide, and 90 centimeters long (1 x 2 x 3 feet). It will hold around 2,000 worms.

Bedding should always be kept as damp as a well-wrung sponge.

Get the bedding ready

Worms need damp but not soggy bedding. Their bodies need to stay moist so they can breathe. If worms dry out, they die. If their bedding is too wet, they will drown.

Worms can survive in many types of bedding. You can use black topsoil or a mixture of shredded cardboard and peat moss. Other materials such as grass clippings, dried leaves, animal manure, and straw can all be mixed in with the bedding.

Ready to recycle!

Empty your worms onto the surface of the bedding in the bin. Put the lid on and let your worms begin recycling! Feed your worms once or twice a week and check their bedding to make sure it remains moist.

Feeding your worms

Make a shallow hole in the bedding and place your kitchen scraps at the bottom. The scraps should be finely chopped by hand or in a food processor before they are fed to the worms because worms cannot eat large chunks of food. Eggshells contain special nutrients that worms need. Add some at least once a week. Cover the waste with bedding.

Crush dried-out eggshells with a rolling pin and chop your vegetables into small pieces before you put them into the bin.

Keeping track

Put your waste in a different spot each time. The diagram on this page will show you one way to keep track of where the food is placed. Follow the same pattern each time and you will not forget. You can also keep track of the temperature in the bin and the amount of food and water you have given your worms by recording these facts in a "worm journal."

In your journal, keep track of the temperature of the bedding, how much you have fed your worms, when you last added water, and where you have placed the food. Use the diagram above as a guide.

Worm math

You need to feed your worms once or twice a week. Each day, one worm eats approximately half the amount of its body weight. If you have 2,000 worms that, all together, weigh one kilogram (2.2 pounds), how much garbage will they eat in one day? What amount would they eat in one week? How much would you feed them if you gave them leftovers twice a week?

Yummy leftovers!

If you were a worm, you would squiggle for joy when you were fed any of the leftovers pictured on this page. Although a worm will eat meat, cheese, and other animal products, it is best to avoid putting these into your bin. Such foods often start to smell as they decay and can attract rats.

A worm's menu

Worms will eat the following leftover foods: apple cores, baked beans, banana peels, cookies, cake, celery tops, cereal, eggshells, grapefruit rinds, onion peels, pineapple, pizza crust, bread, potato peelings, tea bags, coffee grounds, coffee filters, cabbage, lettuce, spinach, macaroni, grapes, watermelon, eggs, and peaches. Worms eat all fruit and vegetable peelings, leaves, grass, and plant cuttings.

Sharing the feast

Worms are not the only creatures that are eating the leftover food scraps in your worm bin. Worms share their meals with many kinds of bugs. If you pick up a handful of vermi-compost and look through it with a toothpick, you will see several kinds of small insects. Some of the creatures are so tiny that they can only be seen with a microscope.

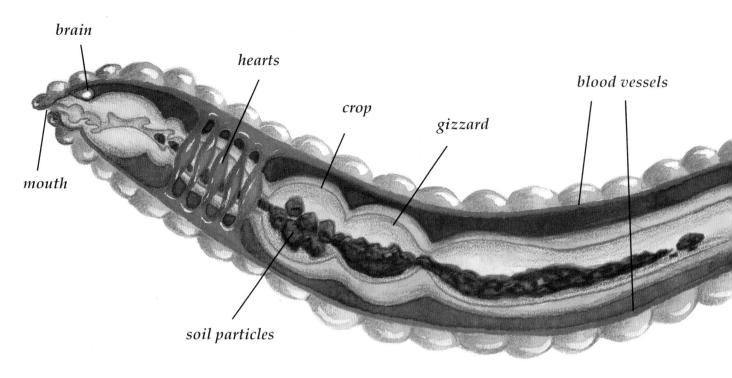

brain

hearts

crop

gizzard

blood vessels

mouth

soil particles

How a worm eats

A worm has a small pad of flesh that sticks out above its mouth. When the worm is hungry, the pad stretches out searching for food. When the worm finds something to eat, the pad pulls the food into the mouth and closes over the mouth. The food has been softened by moisture and by the tiny insects that live in the bin.

The pad stretches out to look for food and pulls it into the mouth underneath.

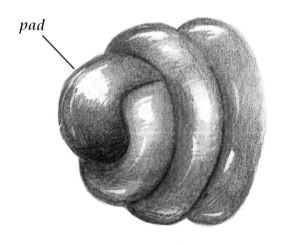

pad

Do worms have teeth?

Worms do not chew food the way you do. They do not have any teeth! They must, instead, grind their food into tiny pieces in a part of their body called the **gizzard**. Hard particles such as sand collect in the worm's gizzard. These particles rub against each other and help grind the food into small pieces.

mouth

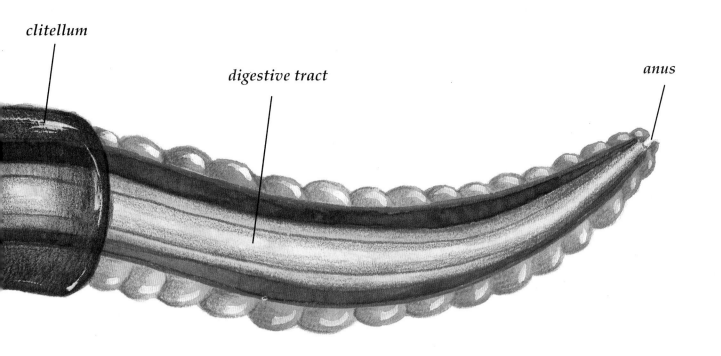

clitellum

digestive tract

anus

Breaking down the food

After the food is ground up, it leaves the gizzard and passes into the worm's **digestive tract**. The digestive tract is a canal that moves the food along and breaks it into tiny food bits. Strong juices in the digestive tract break down the food bits into even smaller pieces.

After the food passes through the digestive tract, the nutritious particles are absorbed into the worm's bloodstream. Five hearts pump the blood throughout the worm's body and carry the food particles to where they are needed. Food that is not nutritious is passed out through the worm's anus. This waste material is called a **casting**.

Worms drag leaves underground so they will rot more quickly. After the leaves break down, the worms eat some of them. The uneaten leaves add nutrition to the soil.

Castings help plants grow

The next time you are relaxing on some green grass, search for tiny castings. Castings look like nubby bits of earth. They are odorless and harmless to people. If you find some, you will know that a worm has been there before you. Pick up one from the ground and crumble it in your hand. The dust that remains in your palm makes up part of the soil on which you are standing.

Farmers love worms!

For thousands of years, worms have been friends to farmers and gardeners. Worms help plants grow and stay healthy. Their tunnels make it possible for air and water to get into the soil and reach the roots of plants.

The drawing below shows the tiny tunnels made by worms through which water and air can reach the roots of plants and trees. How many worms can you find in the picture?

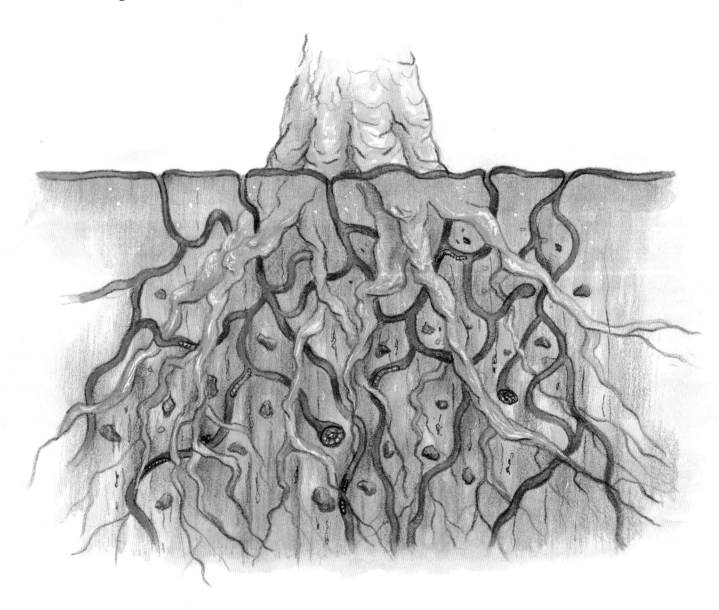

Nutrition for plants

As they dig, worms loosen and mix soil. Their digging allows seeds to sprout and roots to develop. The castings of worms are full of nutrition. They contain the food found in dying plants. This food provides new plants with the energy they need for growing.

Very good friends

The fictional story on the next two pages shows how essential worms are to the environment. Soo, a forest pixie, found out the hard way that life without worms could be very difficult! Read the story and discover why worms are so important to all living beings. Then thank your worms for eating garbage and for the job they perform in nature. When you realize that you, too, are a part of nature, you will know, just as Soo did, that worms are your very good friends!

What would it feel like to be a worm? These students formed a giant worm on the floor of their school and said it felt fine!

The pixie that put the worms to sleep

One day, a foolish young pixie named Soo sprinkled "dozey" dust over the soil of Troll Mountain. As a prank, thinking it would do no harm, the mischievous little pixie caused all the worms to fall into a deep sleep.

The old troll of the forest, who was Soo's teacher, was very dismayed. He asked Soo why she would do such a thing! "I have no use for the lowly earthworm," replied Soo in her usual flippant manner.

The troll shook his head slowly and said, "You have not heeded my lessons well. It is time you learned that all things, even the lowly earthworm, have important roles to play in our forest." And having said that, he turned Soo into an owl and warned her that she must stay that way until she discovered the value of worms.

Soo was not at all unhappy as an owl. In fact, she soon started enjoying herself. Gliding through the forest on silent wings, Soo was aware of all the sounds and movements below her. She was even more certain that a mighty owl such as she did not need to give a hoot about worms!

In time, Soo became very lonely. She did not know any other owls. Then, one day, she heard another owl calling to her. It was a majestic male owl, who took Soo for a mate. Soon, Soo became a mother to three young owlets and was very happy for a time.

As the owlets grew, Soo discovered that she had to work harder and harder to find mice to feed her young. One day, she could find none at all! With her keen owl eyes and senses, Soo set out to learn what had happened to the mice. It did not take her long to discover that there were no mice because there was not enough food for them to eat.

There seemed to be nothing but dead plants and leaves everywhere! Instead of breaking apart and blending into the soil, they were just lying where they fell, covering the surface of the ground. New plants were not able to grow because there was not enough nutrition in the soil. Mice could not find tender plants or seeds to eat, so no baby mice were born!

Soo became very frightened for her owlets and, in a panic, flew through the forest in search of the old troll. When she spotted him, she cried out. "Oh, troll of the forest! The old plants are not coming apart, and new plants are not growing. There are not enough mice, and my owlets are starving!"

"Ah," said the old troll. "I see that you have at last discovered the value of the earthworm."

"What does the problem with the plants have to do with worms?" screeched Soo at the top of her lungs.

The wise old troll answered in a kind and patient voice. "Anything in the forest that is no longer growing must come apart so that the same material may be used over and over to enrich the soil."

"So it is the worm's place to do the taking-apart!" cried Soo, suddenly realizing the importance of the slimy creature that recycled dead plants and added nutrition to the soil. She now understood how the worm helped all the creatures in the forest—even her.

The old troll was pleased with his student and offered to turn her back into a pixie. Soo thought for a moment and agreed to become a pixie again—but only long enough to sprinkle the antidote to dozey dust on the soil. She told her teacher that she wished to remain an owl, raise her owlets, and be part of the forest, just like her new friends—the worms.

Harvesting the vermi-compost

Harvesting the vermi-compost on a sunny day can be lots of fun. Eloise collects the worms she has separated in a margarine container.

Several times a year, you will need to give your worms fresh bedding. In two to three months, worms will eat through their bedding and build up a large amount of castings. If the vermi-compost in the bin has too many castings in it, the worms will get sick.

When it is time to remove the vermi-compost from the worm bin, you will need to set aside a couple of hours for the job. The castings are a valuable part of the vermi-compost—they make nutritious food for plants. Before you change the bedding in your bin, you must first "harvest" the vermi-compost. In order to do this, you need to remove the worms.

Separating the worms

1. Dump the contents of the worm bin onto a large plastic sheet.

2. Do your sorting outdoors on a sunny day or shine a bright light on the big pile. Separate the big pile into a number of little piles.

3. Brush some of the vermi-compost off the top and sides of the first pile. The worms will slowly move to the center of each pile to avoid the bright light. Do the same for each pile until all the vermi-compost is removed.

4. After you have separated the vermi-compost, you will find a bunch of worms at the center of each pile. Put them into a clean container.

5. Scoop the vermi-compost off the plastic sheet and place it in a bag. You can now use it to nourish your garden or house plants.

6. Fill your bin with fresh bedding and gently pour the worms onto the surface. Your worms are again ready to wage war against garbage!

Letting the worms do it

If you do not want to go to the trouble of separating the worms from the vermi-compost, you can remove half the vermi-compost and add new bedding. Shine a bright light over the open bin. The worms will go deep down into the vermi-compost. Now, carefully skim off layers of vermi-compost until you are halfway down. Add new bedding and repeat the procedure every few weeks.

Dump the vermi-compost.

Separate into piles.

Brush vermi-compost away from worms.

Save the vermi-compost for your garden.

Collect worms and put them into new bedding.

The clitellum makes mucus. The sperm from one worm passes through the mucus to the other worm.

Babies, babies, babies!

You are either a boy or a girl, but worms are both male and female at the same time! A worm that has a band, called a **clitellum**, around its body is able to make babies.

To make babies, worms meet and trade their male sperm. They then separate.

Each worm develops a hard substance around its body called a **cocoon**. The worm leaves its own eggs and some of the sperm it collected from its mate in the lemon-shaped cocoon, which is about the size of a kernel of popping corn. The worm then backs out of the cocoon and leaves it behind.

A cocoon forms on the clitellum of each worm.

The cocoon comes off the worm over its head. As the worm backs out of the cocoon, the cocoon picks up eggs and sperm from special openings in the worm's skin.

Can you spot the cocoons on the plate? There are also adult worms, baby worms, and castings. Point to each in the picture.

Cocoons look like tiny lemons.

Inside the cocoon, baby worms are formed from the eggs of one worm and the sperm of another. After three weeks, two or three tiny worms will wiggle out of the cocoon.

Baby worms wiggle out of their cocoon.

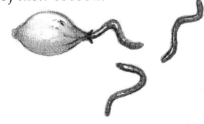

Thousands of baby worms!

Fertilization takes place inside the cocoon. Fertilization is the joining of a sperm with an egg. In about three weeks, two or three baby worms will hatch out of the cocoon. Under ideal conditions, eight adult worms could produce 1,500 babies in six months, so you may need to give some worms away when you change the bedding.

Worm cocoons

Baby worms spend their first three weeks inside the cocoon. When the worms leave the cocoon, they are just a few millimeters long and a whitish color. Baby worms are not tiny for long, however. In about six weeks, they are fully grown and can also begin laying cocoons. If you look carefully, you will see many cocoons in your bin.

Vermi-composting at school

The students pictured on these two pages have a vermi-composter in their classroom. They feed their lunch leftovers to the worms each day. The worms eat the garbage and help the students learn about nature.

Rebecca and Ian are weighing the lunch leftovers of their class, which have been chopped up in a food processor. They are planning to find out how many kilograms of waste their whole school creates in one year. How much food waste does your school produce in a day, week, month, and year?

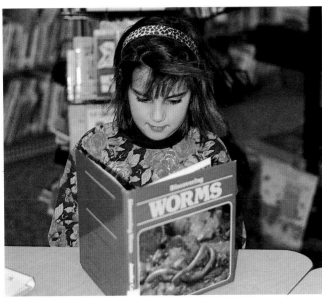

The students in the top and bottom left pictures are learning about the life cycle of the Red Wiggler. They are preparing a project for the school science fair. All three are enjoying their research. The students want to find out everything they can about worms because they share their classroom with them.

The boys decided to share their knowledge with others by writing a newsletter about worms. They are checking a few facts before they type their final draft on the computer. They want to make sure their information is correct. The newsletter will be distributed throughout the school.

A worm talk show

The class of the students shown on these pages held a worm talk show. To prepare for their show, the students wrote questions about worms, such as the ones found on the following pages. With the help of their teacher and librarian, they found the answers. They then learned the answers to everyone else's questions. On the day of the show, each student asked a question and one of the worm puppets answered. Some of the worm puppets tried to wiggle out of answering, but fellow worms came to the rescue. A few talented worms recited poems and sang songs. In the end, all the class members squirmed their way into the spotlight. After the show, everyone was invited to a "casting" party, where all kinds of leftovers were served!

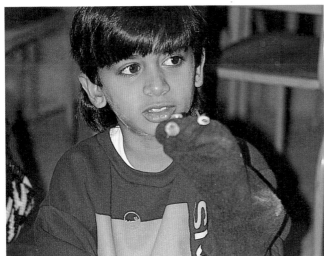

(top) The whole class did a worm salute.
(bottom) Each worm was made from a sock.

These theatrical worms put on a little play
for their friends, who squiggled with joy.

Wondering about worms

Are worms fuzzy wuzzy?

Do you know the song "Nobody Likes Me?" Now that worms are your pets, you may feel like singing some new words to this song. After you have read both the old and new versions of this song, write your own song and make up a dance about worms.

Old version:

Nobody likes me
Everybody hates me
I'm going to the garden to eat worms!
Great, big, juicy ones,
Long, thin, skinny ones,
Itsy-bitsy, fuzzy-wuzzy worms
Yum! Yum!

New version:

Everybody likes me
Nobody hates me
Because they know that I feed worms!
Great, big, juicy ones,
Long, thin, skinny ones,
Itsy-bitsy, fuzzy-wuzzy worms
They're great!

Hairy squirmers

In the song "Nobody Likes Me," worms are called "fuzzy wuzzy." Although you cannot see any fuzz, worms really are covered with hair! Bristly hairs help worms move by gripping the ground.

When worms are in tunnels, they cling so well using their hairs that birds have trouble pulling them out of the soil. The hairs are connected to muscles inside the worm. The worm begins to move by stretching out its head. It then pulls along each section of its body.

Do worms have feelings?

Worms do not have feelings such as happiness or sadness, but they do have senses. They can sense differences in light and touch. They like to eat certain foods more than others. Worms are also sensitive to the amount of water in the soil.

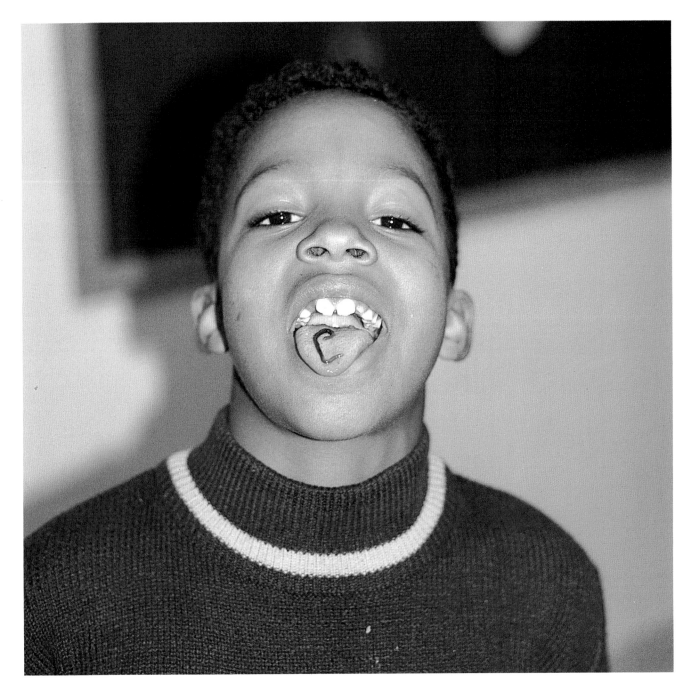

Worms for lunch anyone?

There is a long list of creatures that find worms a tasty meal. Birds, moles, foxes, toads, snakes, beetles, slugs, and flatworms all eat earthworms. If you were ever lost in the wilderness, you could even survive by eating worms!

The boy in the picture above shows that worms do not taste all that bad. Don't worry, though! He did not really eat the worm.

How do worms breathe?

All animals, including worms, need oxygen to stay alive. Oxygen is an odorless and colorless gas that is part of the air we breathe. Oxygen is also a part of water. Worms breathe by taking in oxygen through the wet surfaces of their bodies. The oxygen moves to the worm's blood and is used by the worm's organs.

Do worms have a brain?

A worm's brain is not like an animal or human brain. The worm has a large nerve that runs along its body and a clump of nerves that ends in its head. These nerves control the worm's actions.

One worm or two?

No worm that is cut in half will ever survive to form two healthy worms. There are, however, some types of worms that can grow new tails if theirs have been cut off.

Stink or no stink?

The Red Wiggler has a Latin name (Eisenia foetida) which means "stinker." It is named stinker because of the odor of the manure in which it lives. Manure piles are the Red Wiggler's natural habitat. If they are properly cared for, Red Wigglers and their bedding have almost no odor in a vermi-composting unit.

How long do worms live?

In their natural habitats, worms do not live longer than a year, but worms that are properly cared for in a worm bin can live as long as four years!

Castings for cat litter?

Some people use worm castings as cat litter. Castings are dry and odorless and can easily absorb moisture and reduce unpleasant smells.

Inside or outside?

The Red Wigglers used for vermi-composting cannot be left outside in colder months because they will freeze and die. They should be taken inside before the weather gets too chilly.

If you have raccoons living around your home, they may find your worms a tasty treat! To prevent a raccoon raid, make sure you put your worm bin in a safe place.

How wormy is your world?

Just how many worms are on your front lawn or in the park nearby? The next time it rains, take a flashlight and go outside at night with your parents. Shine the light on the grass and you will see earthworms everywhere! If you are careful not to make any noise with your feet, you may be able to catch a worm before it quickly slips back into the ground.

Do worms have eyes?

When people draw pictures of worms, the worms often have eyes and noses. Real worms have neither!

Will dew worms do?

The big, fat dew worms that people use for fishing are not good for vermi-composting. Dew worms need a lot of room. They cannot survive in the cramped quarters of a vermi-composting bin.

Where do dead worms go?

Worms that die in the bin are eaten by the insects that share the bin with them.

Do I need a worm-sitter?

If you have a vermi-composter, you may have to find a home for your worms during the holidays. Worms can be left alone for several days, but if you take a long vacation, you should make arrangements for a worm-sitter. Your worms need food, and their bedding must stay moist.

The joy of vermi-composting

Now that you have brought worms into your home, what have you learned about nature's ways? How do you feel about worms? With a group of vermi-composting buddies, write songs and poems about the joy of vermi-composting. Have fun getting your neighbors and friends involved! It does not matter how young or old they are, they will discover the joy of vermi-composting!

Glossary

anterior Situated near or toward the front

anus The opening at the end of the digestive tract through which castings leave the worm's body

casting The waste "cast off" from worms

clitellum The band around a worm's body that shows a worm is ready to make babies

cocoon A hard case in which baby worms grow

compost A mixture of decaying organic matter used as fertilizer

decay To rot

environment The surroundings that affect the existence of living beings

fertilization The beginning of reproduction; the union of sperm with an egg

gizzard A part of the worm's stomach that crushes food

habitat The natural surroundings of an animal or plant

harvest To gather

landfill site A low area of land in which garbage and soil are placed in layers

mate To unite as a couple for producing offspring

mucus A thick, slimy fluid

nerve A bundle of fibers that connect the central nervous system to organs and body parts

nutrient Anything that nourishes

organ A part of the body with a special purpose

oxygen An odorless, colorless gas that is essential for supporting life

peat moss A moss used to enrich soil in gardening

posterior Situated near or toward the rear

reproduce To create offspring

scavenger An animal that feeds on dead or decaying animals

sperm The male reproductive cell

topsoil The surface layer of soil

Index

2 3 4 5 6 7 8 9 0 Printed in USA 1 0 9 8 7 6 5 4 3